YEAR 6

STAR MATHS STARTERS

A fresh approach to mental maths

TERMS AND CONDITIONS

IMPORTANT – PERMITTED USE AND WARNINGS – READ CAREFULLY BEFORE USING

Minimum specification:
- PC with a CD-ROM drive and 512 Mb RAM (recommended)
- Windows 98SE or above/Mac OSX.1 or above
- Recommended minimum processor speed: 1 GHz
- Facilities for printing

Julie Cogill and Anthony David

Authors
Julie Cogill and Anthony David

Anthony David dedicates this book to his wife, Peachey, and sons Oliver and Samuel.

Editors
Niamh O'Carroll and Carolyn Richardson

Assistant Editor
Pollyanna Poulter

Illustrator
Theresa Tibbetts (Beehive Illustration)

Series Designer
Joy Monkhouse

Designers
Rebecca Male and Shelley Best

Text © 2008 Julie Cogill and Anthony David
© 2008 Scholastic Ltd

CD-ROM development in association with Vivid Interactive

Designed using Adobe InDesign and Adobe Illustrator

Published by Scholastic Ltd
Villiers House, Clarendon Avenue,
Leamington Spa, Warwickshire CV32 5PR
www.scholastic.co.uk

Printed by Tien Wah, Singapore
1 2 3 4 5 6 7 8 9   8 9 0 1 2 3 4 5 6 7

ISBN 978-1407-10012-8

ACKNOWLEDGEMENTS
Extracts from the Primary National Strategy's *Primary Framework for Mathematics* (2006)
www.standards.dfes.gov.uk/primaryframework, *Renewing the Primary Framework* (2006) and the
Interactive Teaching Programs originally developed for the National Numeracy Strategy © Crown
copyright. Reproduced under the terms of the Click Use Licence.

Every effort has been made to trace copyright holders for the works reproduced in this book, and the
publishers apologise for any inadvertent omissions.

British Library Cataloguing-in-Publication Data
A catalogue record for this book is available from the British Library.

Introduction

In the 1999 *Framework for Teaching Mathematics* the first part of the daily mathematics lesson is described as 'whole-class work to rehearse, sharpen and develop mental and oral skills'. The Framework identified a number of short, focused activities that might form part of this oral and mental work. Teachers responded very positively to these 'starters' and they were often judged by Ofsted to be the strongest part of mathematics lessons.

However, the renewed *Primary Framework for Mathematics* (2006) highlights that the initial focus of 'starters', as rehearsing mental and oral skills, has expanded to become a vehicle for teaching a range of mathematics. 'Too often the "starter" has become an activity extended beyond the recommended five to ten minutes' (*Renewing the Primary Framework for mathematics: Guidance paper,* 2006). The renewed Framework also suggests that 'the focus on oral and mental calculation has been lost and needs to be reinvigorated'.

Star Maths Starters aims to 'freshen up' the oral and mental starter by providing focused activities that help to secure children's knowledge and sharpen their oral and mental skills. It is a new series, designed to provide classes and teachers with a bank of stimulating interactive whiteboard resources for use as starter activities. Each of the 30 starters offers a short, focused activity designed for the first five to ten minutes of the daily mathematics lesson. Equally, the starters can be used as stand-alone oral and mental maths 'games' to get the most from a spare ten minutes in the day.

About the book

Each book includes a bank of teachers' notes linked to the interactive whole-class activities on the CD-ROM. A range of additional support is also provided, including planning grids, classroom resources, generic support for using the interactive whiteboard in mathematics lessons, and an objectives grid.

Objectives grid

A comprehensive two-page planning grid identifies links to the *Primary Framework for Mathematics* strands and objectives. The grid also identifies one of six starter types, appropriate to each interactive activity (see page 7 for further information).

Starter Number	Star Starter Title	Page No.	Strand	Learning objective as taken from the Primary Framework for Mathematics	Type of Starter
16	Maths Boggle: decimals	28	Calculating	Use efficient written methods to add and subtract integers and decimals	Refine
17	Target number: multiplication	29	Calculating	Use efficient written methods to multiply two-digit and three-digit integers by a two-digit integer	Rehearse
18	Finding reflections	30	Understanding shape	Visualise and draw on grids where a shape will be after reflection	Recall
19	Maps and directions	31	Understanding shape	Visualise and draw on grids where a shape will be after translation and rotation	Rehearse
20	Finding rotations	32	Understanding shape	Visualise and draw on grids where a shape will be after rotation through 90° or 180° about its centre or one of its vertices	Refine
21	Coordinates (ITP)	33	Understanding shape	Use coordinates in the first quadrant to locate shapes	Refine
22	Fixing points (ITP)	34	Understanding shape	Estimate angles and use a protractor to measure them	Refine
23	Calculating angles (ITP)	35	Understanding shape	Calculate angles in a triangle or around a point	Refine
24	Convert the weight	36	Measuring	Select and use standard metric units of measure and convert between units using decimals to two places	Refine

Highlighted text indicates the end-of-year objectives

Activity pages

Each page of teachers' notes includes:

Learning objectives
Covering the strands and objectives of the renewed *Primary Framework for Mathematics*

Type of starter
Identifying one or more of the 'six Rs' of oral and mental work (see page 7)

Whiteboard tools
Identifying the key functions of the accompanying CD-ROM activity

What to do
Outline notes on how to administer the activity with the whole class

Differentiation
Adapting the activity for more or less confident learners

Key questions
Probing questions to stimulate and sustain the oral and mental work

Annotations
At-a-glance instructions for using the CD-ROM activity.

Whiteboard hints and tips

Each title offers some general support identifying practical mathematical activities that can be performed on any interactive whiteboard (see pages 8-9).

Recording sheets

Two recording sheets have been included to support your planning:
- Planning for the six Rs: plan a balance of activities across the six Rs of mental and oral maths (see page 7).
- Star Maths Starters diary: build a record of the starters used; titles, objectives covered, how they were used and the dates they were used (see page 47).

About the CD-ROM

Types of activity

Each CD-ROM contains 30 interactive starter activities for use on any interactive whiteboard. These include:

Interactive whiteboard resources
A set of engaging interactive activities specifically designed for *Star Maths Starters*. The teachers' notes on pages 13-42 of this book explain how each activity can be used for a ten-minute mental maths starter, with annotated screen shots giving you at-a-glance support. Similarly, a 'what to do' function within each activity provides at-the-board support.

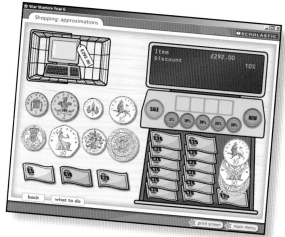

Interactive teaching programs (ITPs)

A small number of ITPs, originally developed by the National Numeracy Strategy, has been included on each CD-ROM. They are simple programs that model a range of objectives, such as data presentation or fraction bars. Their strength is that they are easy to read and use. If you press the 'Esc' button the ITP will reduce to a window on the computer screen. It can then be enlarged or more ITPs can be launched and set up to model further objectives, or simply to extend the objective from that starter. To view the relevant 'what to do' notes once an ITP is open, press the 'Esc' button to gain access to the function on the opening screen of the activity.

Interactive 'notepad'

A pop-up 'notepad' is built into a variety of activities. This allows the user to write answers or keep a record of workings out, and includes 'pen', 'eraser' and 'clear' tools.

Teacher zone

This teachers' section includes links from the interactive activities to the *Primary Framework for Mathematics* strands, together with editable objectives grids, planning grids and printable versions of the activity sheets on pages 43–46.

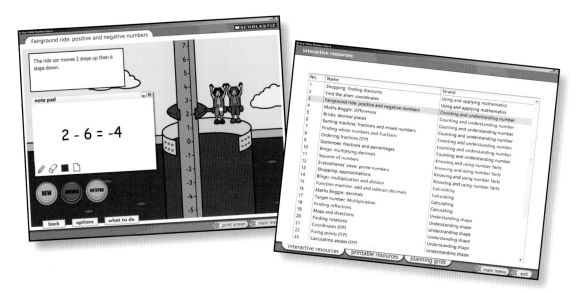

How to use the CD-ROM

System requirements

Minimum specification
- PC with a CD-ROM drive and 512 Mb RAM (recommended)
- Windows 98SE or above/Mac OSX.1 or above
- Recommended minimum processor speed: 1 GHz

Getting started

The *Star Maths Starters* CD-ROM should auto run when inserted into your CD drive. If it does not, use **My Computer** to browse the contents of the CD-ROM and click on the 'Star Starters' icon.

From the start-up screen you will find four options: select **Credits** to view a list of credits. Click on **Register** to register the product to receive product updates and special offers. Click on **How to use** to access support notes for using the CD-ROM. Finally, if you agree to the terms and conditions, select **Start** to move to the main menu.

For all technical support queries, please phone Scholastic Customer Services on 0845 6039091.

The six Rs of oral and mental work

In the guidance paper *Renewing the Primary Framework for mathematics* (2006), the Primary National Strategy identified six features of children's mathematical learning that oral and mental work can support. The description of the learning and an outline of possible activities are given below:

Six Rs	Learning focus	Possible activities
Rehearse	To practise and consolidate existing skills, usually mental calculation skills, set in a context to involve children in problem solving through the use and application of these skills; use of vocabulary and language of number, properties of shapes or describing and reasoning.	Interpret words such as more, less, sum, altogether, difference, subtract; find missing numbers or missing angles on a straight line; say the number of days in four weeks or the number of 5p coins that make up 35p; describe part-revealed shapes, hidden solids; describe patterns or relationships; explain decisions or why something meets criteria.
Recall	To secure knowledge of facts, usually number facts; build up speed and accuracy; recall quickly names and properties of shapes, units of measure or types of charts, graphs to represent data.	Count on and back in steps of constant size; recite the 6-times table and derive associated division facts; name a shape with five sides or a solid with five flat faces; list properties of cuboids; state units of time and their relationships.
Refresh	To draw on and revisit previous learning; to assess, review and strengthen children's previously acquired knowledge and skills relevant to later learning; return to aspects of mathematics with which the children have had difficulty; draw out key points from learning.	Refresh multiplication facts or properties of shapes and associated vocabulary; find factor pairs for given multiples; return to earlier work on identifying fractional parts of given shapes; locate shapes in a grid as preparation for lesson on coordinates; refer to general cases and identify new cases.
Refine	To sharpen methods and procedures; explain strategies and solutions; extend ideas and develop and deepen the children's knowledge; reinforce their understanding of key concepts; build on earlier learning so that strategies and techniques become more efficient and precise.	Find differences between two two-digit numbers, extend to three-digit numbers to develop skill; find 10% of quantities, then 5% and 20% by halving and doubling; use audible and quiet counting techniques to extend skills; give coordinates of shapes in different orientations to hone concept; review informal calculation strategies.
Read	To use mathematical vocabulary and interpret images, diagrams and symbols correctly; read number sentences and provide equivalents; describe and explain diagrams and features involving scales, tables or graphs; identify shapes from a list of their properties; read and interpret word problems and puzzles; create their own problems and lines of enquiry.	Tell a story using an interactive bar chart; alter the chart for children to retell the story; starting with a number sentence (eg 2 + 11 = 13), children generate and read equivalent statements for 13; read values on scales with different intervals; read information about a shape and eliminate possible shapes; set number sentences in given contexts; read others' results and offer new questions and ideas for enquiry.
Reason	To use and apply acquired knowledge, skills and understanding; make informed choices and decisions, predict and hypothesise; use deductive reasoning to eliminate or conclude; provide examples that satisfy a condition always, sometimes or never and say why.	Sort shapes into groups and give reasons for selection; discuss why alternative methods of calculation work and when to use them; decide what calculation to do in a problem and explain the choice; deduce a solid from a 2D picture; use fractions to express proportions; draw conclusions from given statements to solve puzzles.

Each one of the styles of starter enables children to access different mathematical skills and each has a different outcome, as identified above. A bingo game, for example, provides a good way of rehearsing number facts, whereas a 'scales' activity supports reading skills. In the objectives grid on pages 10–11, the type of each Star Starters activity is identified to make it easier to choose appropriate styles of starter matched to a particular objective. A 'six Rs' recording sheet has also been provided on page 12 (with an editable version on the CD-ROM) to track the types of starter you will be using against the strands of the renewed Framework.

Using the interactive whiteboard in primary mathematics

The interactive whiteboard is an invaluable tool for teaching and learning mathematics. It can be used to demonstrate and model mathematical concepts to the whole class, offering the potential to share children's learning experiences. It gives access to powerful resources - audio, video, images, websites and interactive activities - to discuss, interact with and learn from. *Star Maths Starters* provides 30 quality interactive resources that are easy to set up and use and which help children to improve their mathematical development and thinking skills through their use as short, focused oral and mental starters.

Whiteboard resources and children's learning

There are many reasons why the whiteboard, especially in mathematics, enhances children's learning:
- Using high-quality interactive maths resources will engage children in the process of learning and developing their mathematical thinking skills. Resources, such as maths 'games' can create a real sense of theatre in the whole class and promote a real desire to achieve and succeed in a task.
- As mentioned above, the whiteboard can be used to demonstrate some very important mathematical concepts. For example, many teachers find that children understand place value much faster and more thoroughly through using interactive resources on a whiteboard. Similarly, the whiteboard can support children's visualisation of mathematics, especially for 'shape and space' activities.
- Although mathematics usually has a correct or incorrect answer, there are often several ways of reaching the same result. The whiteboard allows the teacher to demonstrate methods and encourages children to present and compare their own mental or written methods of calculation.

Using a whiteboard in Year 6

An interactive whiteboard can help children to achieve many of the learning objectives appropriate to Year 6. These include:
- using multiplication squares, or drawings of squares, to highlight the pattern of square numbers, prime numbers and to identify factors;
- ordering fractions, decimals or negative numbers on an undivided number line;
- using calculators to demonstrate different functions or to show number patterns, such as sequences of negative numbers;
- using simple software programs to demonstrate the position of shapes after rotation;
- asking children to plot coordinates on prepared grids.

Practical considerations

For the teacher, the whiteboard has the potential to save preparation and classroom time, as well as providing more flexible teaching.

ICT resources for the interactive whiteboard often involve numbers that are randomly generated, so that possible questions or calculations stemming from a single resource may be many and varied. This enables resources to be used for a longer or shorter time period depending on the purpose of the activity and how children's learning is progressing. *Star Maths Starters* includes many activities of this type.

From the very practical point of view of saving teachers' time, particularly in the starter activity, it is often easier to set up mathematics resources more quickly than those for other subjects. Once the software is familiar, preparation time is saved especially when there is need for clear presentation, as in drawing shapes accurately or creating charts and diagrams for 'handling data' activities.

Maths resources on the interactive whiteboard are often flexible and enable differentiation, so that a teacher can access different degrees of difficulty using the same software. Last but not least, whiteboard resources save time writing on the board and software often checks calculations, if required, which enables more time both for teaching and assessing children's understanding.

Using *Star Maths Starters* interactively

Much has been said and written about interactivity in the classroom but it is not always clear what this means. For example, children coming out to the board and ticking a box is not what is meant by 'whole-class interactive teaching and learning'. In mathematics it is about challenging children's ideas so that they develop their own thinking skills and, when appropriate, encouraging them to make connections across different mathematical topics. As a teacher, this means asking suitable questions and encouraging children to explore and discuss their methods of calculation and whether there are alternative ways of achieving the same result. *Star Maths Starters* provides some examples of key questions that could be asked while the activities are being undertaken, together with suggestions for how to engage less confident learners and stretch the more confident.

If you already have some experience in using the whiteboard interactively then we hope the teaching suggestions set out in this book will take you further. What is especially important is the facility the whiteboard provides to share pupils' mathematical learning experiences. This does not mean just asking children to suggest answers, but using the facility of the board to display and discuss ideas so that everyone can share in the learning experience. Obviously, this needs to be in a way that explores and relates the thinking of individuals to the context of the learning that is happening.

In the best whiteboard classrooms, teachers comment that the board provides a shared learning experience between the teacher and the class, in so far as the teacher may sometimes stand aside while children themselves are discussing their own mathematical methods and ideas.

Starter Number	Star Starter Title	Page No.	Strand	Learning objective as taken from the Primary Framework for Mathematics	Type of Starter
1	Shopping: finding discounts	13	Using and applying mathematics	Solve problems involving percentages	Refine
2	Find the alien: coordinates	14	Using and applying mathematics	Tabulate systematically the information in a problem or puzzle; identify and record the steps needed to solve it	Reason
3	Fairground ride: positive and negative numbers	15	Counting and understanding number	Find the difference between a positve and a negative integer in context	Recall
4	Maths Boggle: differences	16	Counting and understanding number	Find the difference between a positive and a negative integer	Refine
5	Bricks: decimal places	17	Counting and understanding number	Use decimal notation for tenths, hundredths and thousandths	Reason
6	Dominoes: fractions and mixed numbers	18	Counting and understanding number	Express a larger whole number as a fraction of a smaller one (eg recognise that eight slices of a five-slice pizza represents $^8/_5$ or $1^3/_5$ pizzas)	Refresh
7	Finding whole numbers and fractions	19	Counting and understanding number	Express a larger whole number as a fraction of a smaller one	Read
8	Ordering fractions (ITP)	20	Counting and understanding number	Order a set of fractions by converting them to fractions with a common denominator	Reason
9	Dominoes: fractions and percentages	21	Counting and understanding number	Find equivalent percentages and fractions	Refresh
10	Bingo: multiplying decimals	22	Knowing and using number facts	Use knowledge of place value and multiplication facts to 10 × 10 to derive related multiplication and division facts involving decimals	Recall
11	Squares of numbers	23	Knowing and using number facts	Use knowledge of multiplication facts to derive quickly squares of numbers to 12 × 12 and the corresponding squares of multiples of 10	Rehearse Refine
12	Eratosthenes' sieve: prime numbers	24	Knowing and using number facts	Recognise that prime numbers have only two factors and identify prime numbers less than 100	Refine
13	Shopping: approximations	25	Knowing and using number facts	Use approximations to estimate and check results	Refresh
14	Bingo: multiplication and division	26	Calculating	Recall quickly multiplication and division facts	Recall
15	Function machine: add and subtract decimals	27	Calculating	Calculate mentally with decimals	Recall

Starter Number	Star Starter Title	Page No.	Strand	Learning objective as taken from the Primary Framework for Mathematics	Type of Starter
16	Maths Boggle: decimals	28	Calculating	Use efficient written methods to add and subtract integers and decimals	Refine
17	Target number: multiplication	29	Calculating	Use efficient written methods to multiply two-digit and three-digit integers by a two-digit integer	Rehearse
18	Finding reflections	30	Understanding shape	Visualise and draw on grids where a shape will be after reflection	Recall
19	Maps and directions	31	Understanding shape	Visualise and draw on grids where a shape will be after translation and rotation	Rehearse
20	Finding rotations	32	Understanding shape	Visualise and draw on grids where a shape will be after rotation through 90° or 180° about its centre or one of its vertices	Refine
21	Coordinates (ITP)	33	Understanding shape	Use coordinates in the first quadrant to locate shapes	Refine
22	Fixing points (ITP)	34	Understanding shape	Estimate angles and use a protractor to measure them	Refine
23	Calculating angles (ITP)	35	Understanding shape	Calculate angles in a triangle or around a point	Refine
24	Convert the weight	36	Measuring	Select and use standard metric units of measure and convert between units using decimals to two places	Read
25	Measuring jug: read scales	37	Measuring	Use standard metric units of measure and convert between units using decimals	Rehearse
26	Calculating perimeter	38	Measuring	Calculate the perimeter of rectilinear shapes	Refine
27	Finding area	39	Measuring	Calculate the area of rectilinear shapes	Refine
28	Area (ITP)	40	Measuring	Estimate the area of an irregular shape by counting squares	Rehearse
29	Data handling (ITP)	41	Handling data	Solve problems by collecting, selecting, processing, presenting and interpreting data, using ICT where appropriate	Read
30	Line graph (ITP)	42	Handling data	Interpret line graphs	Read

Planning for the six Rs of oral and mental work

Oral and mental activity - six Rs	Using and applying mathematics	Counting and understanding number	Knowing and using number facts	Calculating	Understanding shape	Measuring	Handling data
Rehearse			● Squares of numbers	● Target number: multiplication	● Maps and directions	● Measuring jug: read scales ● Area (ITP)	
Recall		● Fairground ride: positive and negative numbers	● Bingo: multiplying decimals	● Bingo: multiplication and division ● Function machine: add and subtract decimals	● Finding reflections		
Refresh		● Dominoes: fractions and mixed numbers ● Dominoes: fractions and percentages	● Shopping: approximations				
Refine	● Shopping: finding discounts	● Maths Boggle: differences	● Squares of numbers ● Eratosthenes' sieve: prime numbers	● Maths Boggle: decimals	● Finding rotations ● Coordinates (ITP) ● Fixing points (ITP) ● Calculating angles (ITP)	● Calculating perimeter ● Finding area	
Read		● Finding whole numbers and fractions				● Convert the weight	● Data handling (ITP) ● Line graph (ITP)
Reason	● Find the alien: coordinates	● Bricks: decimal places ● Ordering fractions (ITP)					

Shopping: finding discounts

Strand

Using and applying mathematics

Learning objective

Solve problems involving percentages

Type of starter

Refine

Whiteboard tools

● Drag one item from the shop and drop into the shopping basket.
● Press 'check-out' to take the basket to the till on the next screen.
● Press one of the pink buttons on the till to select a percentage discount (5%, 10%, 20%, 25% or 30%).
● Drag the appropriate amount of money and drop into the till to pay for the item.
● Press 'sale' to check if the amount of cash given is correct.
● Press 'clear' to start a new sale of the same item. Press 'back to shop' to select a different item.

What to do

This practical activity gives the children practice in finding percentage discounts on an item and in working out the cost of the item after the discount has been taken off. Once the item has been chosen, ask the children to work out the discount and the discounted cost using small whiteboards so that the calculation can be discussed by the whole class. A calculator may be helpful for some calculations. When calculating discounts, if recurring decimals occur then amounts should be rounded to the nearest penny. Bear in mind that all notes and coins can be used more than once. If the amount entered is incorrect, 'No sale – please try again' appears and the child can have another try.

Differentiation

Less confident: in the early stages of this activity, select items with a 10% discount.
More confident: ask the children to round the cost of the item first and estimate the discount and the discounted cost before proceeding with the exact calculation.

Key questions

● *What is the discount?*
● *How much would you need to pay once the discount has been taken off?*

shopping items
Drag one item for purchase to the basket

'sale'
Press to check if the amount paid is correct

'clear'
Press to start new sale

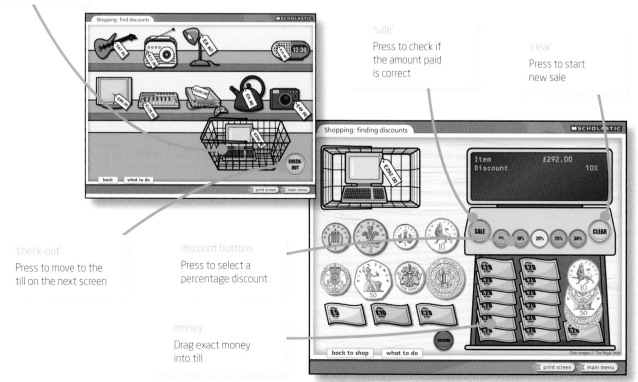

'check-out'
Press to move to the till on the next screen

discount buttons
Press to select a percentage discount

money
Drag exact money into till

Find the alien: coordinates

Strand

Using and applying mathematics

Learning objective

Tabulate systematically the information in a problem or puzzle; identify and record the steps needed to solve it

Type of starter

Reason

Whiteboard tools

● Type the x and y coordinates into the brackets to select a point on the grid.
● A keypad pops up automatically when you press on the brackets to enter a number.
● Press 'check' to confirm the choice.
● Press 'new' to start a new game with the alien in a different position.

What to do

The aim of the activity is to find a point on a 5 × 5 grid at which an alien is hiding. The position of the alien is randomly chosen each time. The child selects, from a given range, the number of degrees he or she wishes to move (left or right, clockwise or anticlockwise) and the number of squares. To avoid this turning into a guessing game, the position of the alien, which is at an intersection of lines, is given as a number of moves from the current position: 'You are n squares away' appears on the screen. The n squares are counted horizontally and/or vertically, not diagonally.

For maximum efficiency in solving the problem, this activity requires a systematic strategy. The children will love doing the activity and will soon develop logical processes, selecting coordinates carefully to find the alien in the fewest goes possible. Once the alien has been found the space picture is completed.

Differentiation

Less confident: to ensure that the less confident children understand the horizontal and vertical numbering system, ask them for coordinates early in the activity – while the choice of grid point is still random.
More confident: ask the children about possible strategies to find the alien in the least number of goes.

Key questions

● *Describe (for example) the top left-hand corner point using horizontal and vertical coordinates.*
● *Now that we know the number of horizontal and vertical moves from where the alien is hiding, at which points might you find him?*

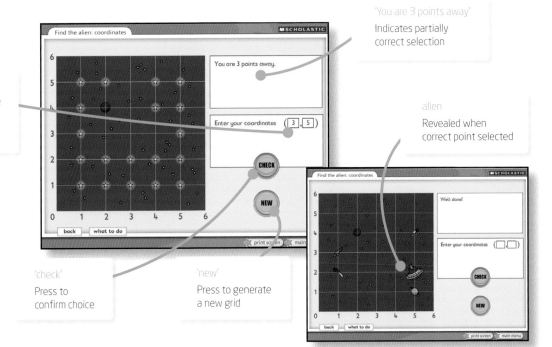

'You are 3 points away'
Indicates partially correct selection

coordinates
● Press to enter here
● Keypad pops up automatically

alien
Revealed when correct point selected

'check'
Press to confirm choice

'new'
Press to generate a new grid

Fairground ride: positive and negative numbers

Strand

Counting and understanding number

Learning objective

Find the difference between a positive and a negative integer in context

Type of starter

Recall

Whiteboard tools

- Start point: the riders start at ground level or zero.
- Move the riders up and down the ride in steps of 1 to solve a maths problem or number sentence.
- Select 'options' to show or hide the related number sentence.
- Press 'new' to move the riders back to starting point and to generate a new question.
- Use the 'notepad' to show calculations.
- Press 'answer' to reveal the answer.

What to do

In this activity, children are encouraged to add positive and negative numbers using mental arithmetic. The screen shows a cross-section of a theme park ride that is partly above and partly below ground; participants ride up and down 'Screamer', at speed. The riders can climb from −12 (below ground) to +12 (above ground). Once the task is set (for example: *The riders go 3 steps up and then 7 steps down*), the children move the riders to the correct position.

Encourage the children to use their individual whiteboards to work out the answers for themselves before any class discussion. The activity may be extended beyond the initial question by asking, for example: *What would happen if the riders now move down another 2 steps... up another 6?* This may be illustrated by dragging and dropping the riders to the new position.

Differentiation

Less confident: support the children by providing them with a copy of photocopiable page 43, 'Fairground ride'.
More confident: ask the children to imagine that the ride went below -12 level. Set some questions that would take the riders from above ground to below -12.

Key questions

- *On what number do the riders finish?*
- *How far up or down would the riders need to move to reach* (for example) *the 5 mark? How many steps to move to -7?*

'answer'
Press to reveal

'new'
- Press to generate question
- Moves the riders back to zero

'notepad'
Press to write calculations

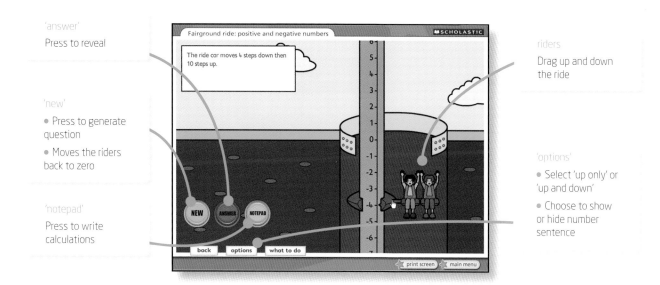

riders
Drag up and down the ride

'options'
- Select 'up only' or 'up and down'
- Choose to show or hide number sentence

Maths Boggle: differences

Strand

Counting and
understanding number

Learning objective

Find the difference
between a positive and a
negative integer

Type of starter

Refine

Whiteboard tools

● Press 'new' to rattle the Boggle dice.
● Change the target by selecting a new question from the 'options' menu.
● Highlight each dice by pressing it once (to remove the highlight, press again).
● Press 'new' for a new set of numbers.
● Use the 'notepad' to show calculations.

What to do

The aim of this activity is to work out differences between positive and negative integers on the grid.

To begin the game, select a question from the 'options' menu or, if you prefer, prepare your own question. Once this has been understood, the dice are 'rattled' to reveal a random selection of numbers. In pairs, or individually, the children set out to find the answer by working out the differences. Challenge individual children to come to the board to write their calculations on the on-screen notepad. Highlight particular dice by pressing on them, for demonstration purposes.

Differentiation

Less confident: identify pairs of integers that have the largest differences. Suggest that the children round the numbers before estimating and calculating the actual differences.
More confident: extend the activity by asking the children to total strings (rows or columns) of positive and negative integers.

Key questions

● *Find the difference between ___ and ___ .*
● *How did you work out the difference? What methods did you use to check your answers?*

Boggle © 2007, Hasbro. All rights reserved.

Bricks: decimal places

Strand

Counting and understanding number

Learning objective

Use decimal notation for tenths, hundredths and thousandths

Type of starter

Reason

Whiteboard tools

● Use 'options' to fix the first digit of the numbers, if required.
● Press 'go' to generate five bricks each with a number between 0.001 and 9.999.
● Drag each brick into the gaps in the wall, with the smallest number in the lowest position to complete the wall.
● If all five bricks are positioned correctly, a 'Well done' message appears. If any bricks are placed incorrectly, a 'Try again' message appears. Press 'ok' and the bricks move back to the starting position.
● Press 'go' again to select a new set of bricks.

What to do

Use this activity either to rehearse existing strategies for ordering three-place decimals or to probe children's reasoning. Press 'go' to reveal five bricks, each showing a number between 0 and 10 with three decimal places. Ask the children to work as a whole class to decide the correct order by writing the answers on their individual whiteboards. Position the bricks in the wall by dragging and dropping them, or ask individual children to come to the board to place them.

Differentiation

Less confident: use an empty number line to support the children's ordering skills, or fix the first digit to limit the number range.
More confident: ask the children what would need to be added to the top brick to make 10.

Key questions

● *What does the number to the left of the decimal point represent on each brick? What does the number to the right of the decimal point represent on each brick?*
● *What would the new number be if 1, 2 or 3 is added to the number on the lowest brick (or 0.1, 0.2 or 0.3 is added to the number on the lowest brick)?*

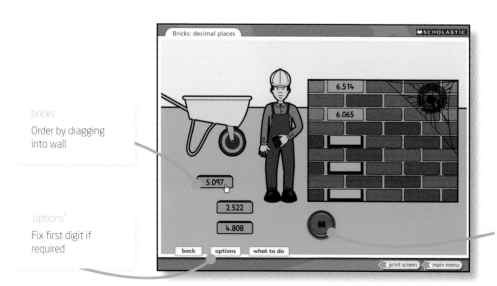

bricks
Order by dragging into wall

'options'
Fix first digit if required

'go'
Press to generate new bricks

Dominoes: fractions and mixed numbers

Whiteboard tools

- Press 'new' to start a new game.
- Press the 'miss a go' button to take another domino from the pot.
- Domino:
  - Drag and drop into the game.
  - Press to rotate 90˚.
- Press 'winner' if Player 1 or Player 2 has placed all of their dominoes.

Learning objective

Express a larger whole number as a fraction of a smaller one

Type of starter

Refresh

What to do

The aim of this activity is to match mixed numbers with their equivalent fractions, and vice versa: for example, $1\frac{3}{4}$ may be matched with $\frac{7}{4}$, or $1\frac{1}{2}$ with $\frac{3}{2}$. The activity is played in the same way as regular dominoes, with two groups playing against each other.

Each group or 'player' (maximum of two) is dealt five dominoes. A starter domino is randomly selected to begin the activity and the players then take it in turns to play. Dominoes are dragged to the playing area; each may be turned through 90° as necessary for correct positioning. If a player is unable to place a domino, he or she must take one from the central pot. This continues until one player has played all his or her dominoes and is declared the winner, or there are no more dominoes in the pot. It is possible to create a stalemate situation, where neither player can play a domino and the pot is empty. In this case, the player with the fewest dominoes is declared the winner.

Please note that there is no automatic checking of whether dominoes are in the correct position – this is left to the agreement of the players.

Differentiation

Less confident: use talk partners to discuss moves, which will help to support children's confidence and affirm their decisions.
More confident: play 'beat the teacher', in which children pit themselves against an adult in the classroom.

Key questions

- *How can we identify which dominoes to use?*
- *Which fraction matches $1\frac{1}{2}$? Are there any other options we might choose?*

players 1 and 2

Panel turns green to indicate whose turn it is

'new'

Press to start new game

'winner'

Press when activity is complete

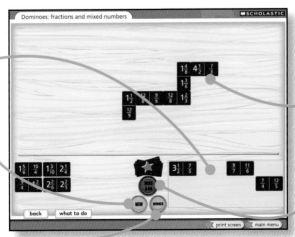

domino

- Drag domino to playing space
- Rotate by pressing top right-hand corner

'miss a go'

Press to take another domino from the pot

Finding whole numbers and fractions

Whiteboard tools

- To change the grid size, press 'options'.
- Press on your chosen highlighter to build up a shape.
- If part of a shape needs to be removed, press on the white highlighter.
- Press 'print screen' to print the images.
- Press 'clear' to start again.

What to do

Build up a line of squares using the red palette (for example, four); then ask the children how many small squares make the whole. Explain that this number is the denominator of the fraction so the four red squares represent $^4/_4$. Next, using different colours, build up several lines of squares below the first that are longer than the red line: these represent improper fractions. For example, a line of six blue squares would represent $^6/_4$ or $1^2/_4$ or $1\frac{1}{2}$. Then ask the children the questions listed below. Once they are confident, operate this process in reverse by choosing a whole number and a fraction first so that you, or the children themselves, can illustrate the fraction.

Differentiation

Less confident: start with easier fractions, such as quarters.
More confident: when appropriate, ask the children if they can cancel any common factors (for example, so that $1\frac{3}{6}$ becomes $1\frac{1}{2}$).

Key questions

- *How many small squares make up the whole? What is this called in the fraction?*
- *What is the numerator of each differently coloured fraction shown?*
- *How can each line of squares be written as a fraction with the same denominator as the red squares, using just fractions or whole numbers and fractions?*

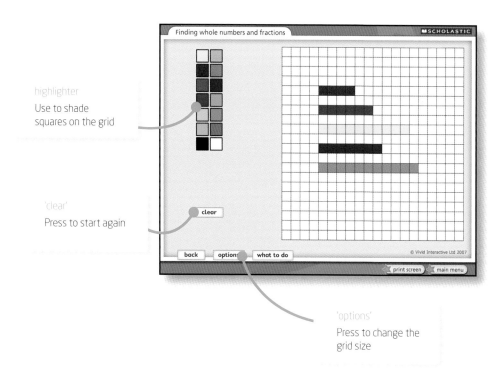

highlighter
Use to shade squares on the grid

'clear'
Press to start again

'options'
Press to change the grid size

Ordering fractions (ITP)

Strand

Counting and understanding number

Learning objective

Order a set of fractions by converting them to fractions with a common denominator

Type of starter

Reason

Whiteboard tools

- Press the small green and yellow bar to produce more fraction bars, up to a maximum of five.
- Press the arrows next to the fraction bar to increase or decrease the denominator, in steps up to 1, and show the fraction chosen.
- Press 'fdpr' to reveal the fractions, decimals, percentages and ratios equivalent to each bar.
- Press the individual fractions on each 'fraction bar' to change the colour from green to yellow to create different fractions on screen.
- Press 'reset' to clear all but the lowest green bar.

What to do

This is a flexible tool that enables equivalent fractions to be easily demonstrated and then ordered. Start with the base fraction bar and build up three more fractions, showing ½, ¼ and ⅛ by changing the denominators to 2, 4 and 8. Any number of different equivalent fractions may be built up in this way to help the children see and understand equivalence. For example, ½, ¾ and ⅝ may easily be compared to show that, starting with the lowest fraction, the order is ½, ⅝ and ¾. Next, show the children how the fractions can all be converted to eighths numerically by multiplying the numerator and denominator of the fraction as appropriate by the same number.

Differentiation

Less confident: start by comparing just two fractions, so that the children understand the principles.
More confident: ask the children to convert all the fractions numerically first, so that they have a common denominator. They should then order them and, finally, check their results using the fraction bars.

Key questions

- *Which do you think is the larger (or largest) of the fractions given?*
- *Can you place the fractions in order of size, starting with the smallest?*

denominator
Press arrows to change denominator

fraction bar
Press to change from green to yellow

add fraction bar
Press to produce up to five fraction bars

Primary *National Strategy*

'fdpr'
Press to show fractions, decimals, percentages and ratios equivalent to each bar

Dominoes: fractions and percentages

Strand

Counting and understanding number

Learning objective

Find equivalent percentages and fractions

Type of starter

Refresh

Whiteboard tools

● Press 'new' to start a new game.
● Press the 'miss a go' button to take another domino from the pot.
● Domino:
  • Drag and drop into the game.
  • Press to rotate 90˚.
● Press 'winner' if Player 1 or Player 2 has placed all of their dominoes.

What to do

The aim of this activity is to match percentages with their equivalent fractions, and vice versa: for example, ¾ may be matched with 75% or ½ with 50%. The activity is played in the same way as regular dominoes, with two groups playing against each other.

Each group or 'player' (maximum of two) is dealt five dominoes. A starter domino is randomly selected to begin the activity and the players then take it in turns to play. Dominoes are dragged to the playing area; each may be turned through 90° as necessary for correct positioning. If a player is unable to place a domino, he or she must take one from the central pot. This continues until one player has played all his or her dominoes and is declared the winner, or there are no more dominoes in the pot. It is possible to create a stalemate situation, where neither player can play a domino and the pot is empty. In this case, the player with the fewest dominoes is declared the winner.

Please note that there is no automatic checking of whether dominoes are in the correct position - this is left to the agreement of the players.

Differentiation

Less confident: use talk partners to discuss moves, which will help to support a child's confidence and affirm their decisions.
More confident: play 'beat the teacher', in which children pit themselves against an adult in the classroom.

Key questions

● *How can we identify which dominoes to use?*
● *What percentage matches the ¹/₅ domino? Which domino would you place next to 25%?*

players 1 and 2

Panel turns green to indicate whose turn it is

'winner'

Press when activity is complete

'new'

Press to start new game

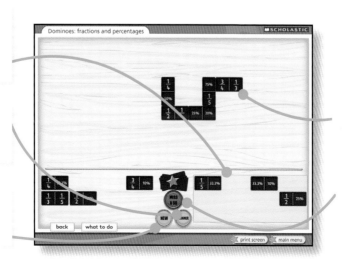

domino

● Drag domino to playing space
● Rotate by pressing top right-hand corner

'miss a go'

Press to take another domino from the pot

Bingo: multiplying decimals

Strand

Knowing and using number facts

Learning objective

Use knowledge of place value and multiplication facts to 10 × 10 to derive related multiplication and division facts involving decimals (eg 0.8 × 7, 4.8 ÷ 6)

Type of starter

Recall

Whiteboard tools

● Press 'set timer' to adjust time between bingo 'calls' (5 to 20 seconds).
● Press 'start' to start a new game.
● Press 'check grid' to check answers if someone calls *House*.
● Press 'play on' or 'winner' after checking a player's grid.

What to do

The aim of this activity is to improve the children's quick recall of multiples of decimals up to 10 × 10, against a time limit. Before the session, print the bingo cards from the CD-ROM, or photocopy the blank bingo cards template on page 44 and make your own. Once the activity has begun, the children have to match the multiplication and division questions, which are displayed or called out, to the answers given on their grid. The children can either play in pairs or individually. If the children are new to the activity, allow for more time between bingo calls; they will become faster and the pace of the activity can increase. A child wins by calling out *House* (or whatever winning call you specify); at this point you should press 'pause' to check whether his or her answers are right. Answer sheets can be checked against the answer 'check grid', which can be called up by pressing the 'check grid' button. If the child's answers are correct, press the 'winner' button for a fanfare!

Differentiation

Less confident: extend the time between each question or ask children to call *House* after correctly pairing only five sets of numbers.
More confident: increase the number of answers on the bingo cards using the bingo card template on photocopiable page 44.

Key questions

● *You know that 0.7 × 9 = 6.3. What division facts can you make from this?*
● *What strategies did you use to remember these facts?*

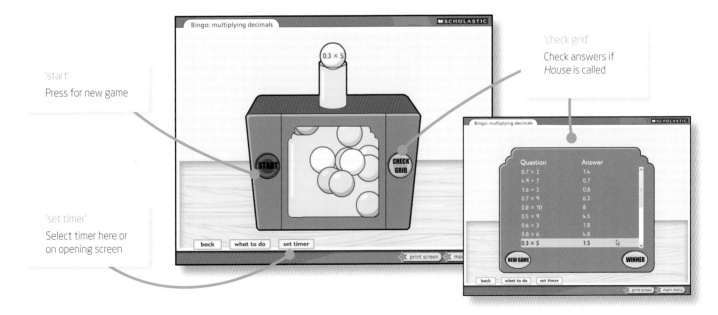

'start'
Press for new game

'set timer'
Select timer here or on opening screen

'check grid'
Check answers if *House* is called

Squares of numbers

Strand

Knowing and using number facts

Learning objective

Use knowledge of multiplication facts to derive quickly squares of numbers to 12 × 12 and the corresponding squares of multiples of 10

Type of starter

Rehearse
Refine

Whiteboard tools

● Press 'options' to select from two multiplication squares: the first shows the multiplication tables from 1 to 12; the second from 10 to 120.
● Use the 'hide' button to hide one of the numbers on the grid.
● Select 'highlight' or 'clear' and then press any hidden square to reveal the number beneath.

What to do

The children's task is to derive all the square numbers in the 12 × 12 table square - using, if necessary, other multiplication facts already visible in the table square. Start by explaining to the children that the square of a number is the number multiplied by itself. Before revealing the square numbers in the table, ask the children (working in pairs) to write down the squares of numbers up to 12. If necessary they may use the table square that is visible on the board. Pressing the square relating to the children's suggestions will confirm or refute their answers. The second task is to repeat this process using the 10 to 120 multiplication square.

Differentiation

Less confident: allow the children to work at their own pace in finding those square numbers that are easier to calculate.
More confident: ask the children if they see any relationship between 3 × 3 and 9 × 9 or between 2 × 2 and 4 × 4.

Key questions

● *What are the squares of all the numbers up to 12?*
● *How could you quickly work out the squares of all the numbers from 10 to 120?*

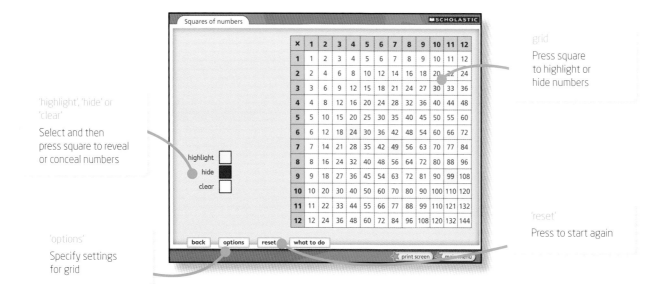

'highlight', 'hide' or 'clear'
Select and then press square to reveal or conceal numbers

'options'
Specify settings for grid

grid
Press square to highlight or hide numbers

'reset'
Press to start again

Eratosthenes' sieve: prime numbers

Strand

Knowing and using number facts

Learning objective

Recognise that prime numbers have only two factors and identify prime numbers less than 100

Type of starter

Refine

Whiteboard tools

● Press 'options' to change the start number and number of squares in the grid.
● Press any square to highlight the number.
● Use the 'hide' button to hide one of the numbers on the grid.
● Select 'highlight' or 'clear' and then press any hidden square to reveal the number beneath.

What to do

Eratosthenes was a Greek mathematician, geographer and astronomer who lived from 276BC to 194BC. He devised this method of quickly finding prime numbers. The number square appears with all numbers showing up to 100. The children's task is to find all the prime numbers by highlighting all those numbers that are *not* prime. Remind them that 1 is not a prime number but 2 is, and start by highlighting the number 1. Then highlight all the multiples of 2, apart from 2. Next, highlight 3 and all subsequent multiples of 3, by using the children's knowledge of the three-times table - or by counting on in threes if necessary. Continue eliminating multiples up to 100. All those numbers that have *not* been highlighted are the prime numbers up to 100. Ask the children to write these down.

Differentiation

Less confident: ask the children to highlight the easier multiples, such as multiples of 2.
More confident: ask the children why it is only necessary to highlight up to multiples of 7.

Key questions

● *Why is it unnecessary to highlight multiples of 4, 6 and 8 once the even numbers have been highlighted?*
● *Why is it unnecessary to highlight multiples of 9 once multiples of 3 have been highlighted?*

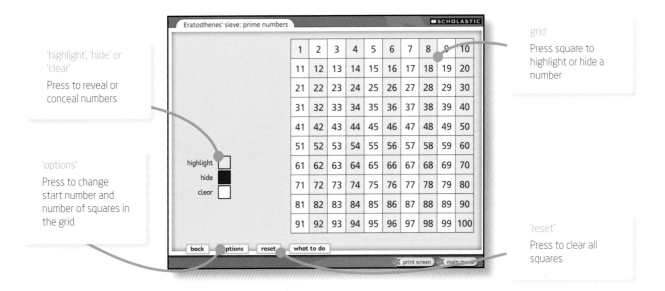

'highlight', 'hide' or 'clear'
Press to reveal or conceal numbers

'options'
Press to change start number and number of squares in the grid

grid
Press square to highlight or hide a number

'reset'
Press to clear all squares

Shopping: approximations

Strand

Knowing and using number facts

Whiteboard tools
- Drag two items from the shop into the shopping basket.
- Press 'check-out' to take the basket to the till on the next screen.
- Drag cash into the till.
- Press the 'sale' button to check if amount is correct.
- If it is incorrect, 'No sale - please try again' appears.
- Press 'back to shop' to start a new sale.

Learning objective

Use approximations to estimate and check results

Type of starter

Refresh

What to do

This activity is designed to encourage the children to round amounts prior to either adding mentally, using two-column addition, or using a calculator, to find totals and then to check their estimates against calculations. Once the items have been chosen, ask the children to work out the answers to the questions below using their individual whiteboards so that the calculations can be discussed by the whole class. Bear in mind that all notes and coins can be used more than once.

Differentiation

Less confident: to start with, ask the children to round amounts to the nearest pound, to give them confidence.
More confident: ask the children to estimate the cost of all the goods in the shop.

Key questions
- *What would each item cost if it were rounded to the nearest £1 or £10? What is the estimated cost of both items?*
- *What coins could you use to pay the exact amount?*

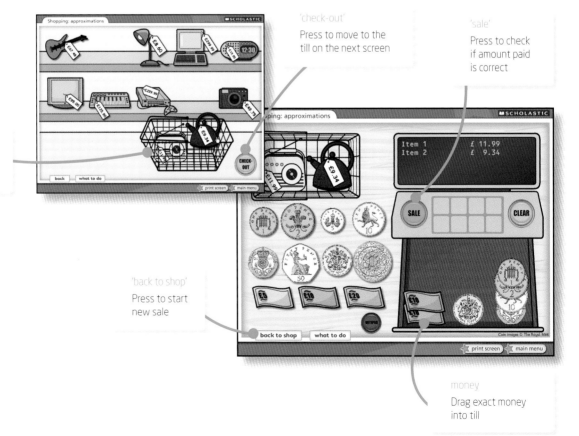

'check-out'
Press to move to the till on the next screen

'sale'
Press to check if amount paid is correct

shopping items
Select and drag two items into the basket

'back to shop'
Press to start new sale

money
Drag exact money into till

Bingo: multiplication and division

Strand

Calculating

Learning objective

Recall quickly multiplication and division facts

Type of starter

Recall

Whiteboard tools

- Press 'set timer' to adjust time between bingo calls (5 to 20 seconds).
- Press the 'start' button to start a new game.
- Press 'check grid' to check answers if someone calls *House*.
- Press 'play on' or 'winner' after checking a player's grid.

What to do

The aim of this activity is to improve the children's quick recall of multiplication and division facts, against a time limit. Before the session, print the bingo cards from the CD-ROM, or photocopy the bingo card template on page 44 and make your own.

Each ball offers a different number sentence. If the answer appears on their bingo card, ask the children to mark it off. If they are new to the activity, allow for more time between bingo calls; they will become faster and the pace of the activity can increase. A child wins by calling out *House* (or whatever winning call you specify); at this point, you should press 'pause' to check whether his or her answers are right. Answer sheets can be checked against the answer grid by pressing the 'check grid' button. If the child's answers are correct, press the 'winner' button for a fanfare!

Differentiation

Less confident: extend the time between each question, or ask children to call *House* after correctly pairing only five sets of numbers.

More confident: increase the number of answers on the bingo cards using the bingo card template on page 44.

Key questions

- *Which questions did you find difficult?*
- *What strategies did you use to remember these facts?*

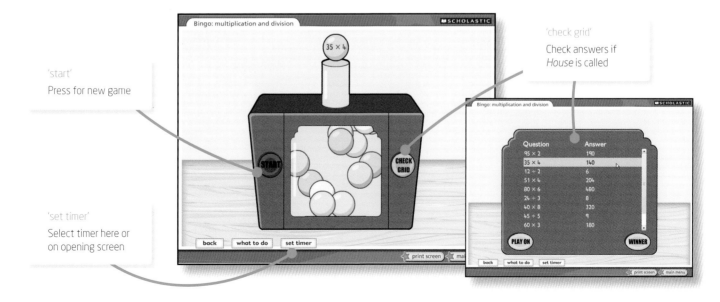

'start'
Press for new game

'set timer'
Select timer here or on opening screen

'check grid'
Check answers if *House* is called

Function machine: add and subtract decimals

Strand

Calculating

Learning objective

Calculate mentally with decimals

Type of starter

Recall

Whiteboard tools
- Use the 'options' menu to set the 'machine mode'. Select from 'manual' or 'random'.
- Select 'manual' to prepare your own number sentence, or 'random' to produce a computer-generated number sentence.
- Use the pop-up keypad, which appears when you press on a window to enter a number.
- Press the 'history' button to view a list of the number sentences completed during the lesson.

What to do
The aim of this activity is to find any missing decimal numbers or the function in order to complete the number sentence. Either you or the computer can generate these using either the 'random' or 'manual' mode options.

Manual mode: prepare some number sentences involving the addition or subtraction of pairs of one- or two-place decimal fractions or a mix of whole numbers to 10 and a decimal fraction – for example, 3.15 + 1.85 = □; 2 – 1.31 = □ and so on. Use the drop-down menu in the function window to select either a + or – operation. Press 'go' to check answers.

Random mode: the computer selects a number sentence, but hides the input, output and function windows on the machine. Decide which element to reveal first and press that window to open it. After one other element has been revealed, ask the children to write down and then display the missing number or function. Check their answers and then press 'go' to check the answer on the machine.

Differentiation
Less confident: work in 'manual' mode and limit the number range to whole numbers and one-place decimals, as required.
More confident: in 'manual' mode, focus upon adding pairs of two-place decimals.

Key questions
- *How can you find the missing parts of the number sentence?*
- *How much do I need to show you before you can complete this number sentence?*
(Progressively reveal different parts of the number sentence.)

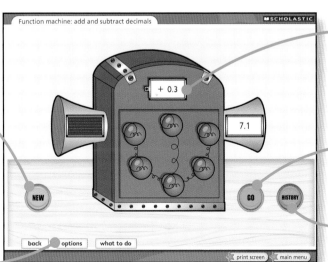

'new'
- Press to start again in 'manual' mode
- Press for number sentence in 'random' mode

'options'
- Select 'manual' mode to enter your own number
- Select 'random' mode for computer-generated numbers, initially hidden

windows
- Type numbers and functions in 'manual' mode
- Press to open in 'random' mode

'go'
Press to check answer

'history'
Press to reveal all number sentences used

Maths Boggle: decimals

Strand

Calculating

Learning objective

Use efficient written methods to add and subtract integers and decimals

Type of starter

Refine

Whiteboard tools

- Press 'new' to rattle the Boggle dice.
- Change the target by selecting a new question from the 'options' menu.
- Highlight each dice by pressing it once (to remove the highlight, press again).
- Press 'new' for a new set of numbers.
- Use the 'notepad' to show calculations.

What to do

The aim of this activity is to use written methods of addition and to reinforce the children's understanding of notation for tenths and hundredths. Where this is well established, encourage the children to use more sophisticated strategies to estimate, then calculate, answers to a set of prepared questions. In this version of the activity, questions relate to strings (rows or columns) of decimal fractions.

To begin the game, select a target question from the 'options' menu or, if you prefer, you may set your own target. Once this has been understood, the dice are rolled to reveal a random selection of numbers. In pairs, or individually, the children set out to answer the questions by adding the numbers on the screen. Invite children to come to the board to write their answers using the on-screen notepad. Check answers by pressing the dice to highlight them.

Differentiation

Less confident: use a limited range of questions – for example, adding selected pairs. Focus on children's written methods of adding the decimal fractions.
More confident: ask children to think of their own questions to ask a partner or set them an additional challenge, such as: *Find pairs of numbers that can be added to make a whole number.*

Key questions

- *How did you work out the highest/lowest total? What written method did you use?*
- *How can you quickly check your answers?*

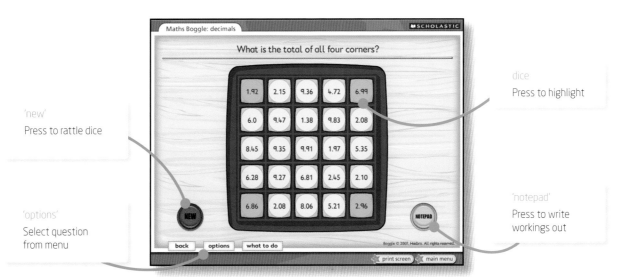

Boggle © 2007, Hasbro. All rights reserved.

Target number: multiplication

Strand
Calculating

Learning objective
Use efficient written methods to multiply two- and three-digit integers by a two-digit integer

Type of starter
Rehearse

Whiteboard tools
- Press 'go' to generate five number cards.
- In 'options' select 'randomly generated' to produce a computer-generated target number, or select 'entered by teacher' to manually insert a target number.
- Use the 'notepad' to work out calculations. A 'pen' tool will automatically pop up when the notepad opens. Press 'start again' to delete any text or use the 'eraser' tool.
- Press 'winner' if the children complete the activity successfully.

What to do
The aim of this activity is to use known number facts to find the target number. The activity provides a random target number and the children are asked to show how this number may be reached using the numbers given. For example, the target may be 2400 and the numbers produced that can be used to find the target may be 20, 40, 100, 200, 1000. Alternatively, you can write your own number by selecting 'options' and 'entered by teacher'.

The key principle behind this activity is that the children use all four operations (+, −, ×, ÷) and use learned strategies in order to write number sentences that match or nearly match the target number. It is the process that is important. Invite individual children to write their calculation on the on-screen notepad, demonstrating that it matches or nearly matches the target. At this point you can address any challenges, or allow the rest of the class to suggest alternative methods. Press 'winner' if the calculation is correct.

Differentiation
Less confident: support the children with written multiplication methods where necessary. For additional support, provide them with the photocopiable 'Target number' sheet on page 45.

More confident: encourage the children to use squared numbers. Ask: *Can you reach the target number more quickly using this method?*

Key questions
- *What strategies are the most efficient? How do these strategies help you to 'hit the target'?*
- *What tips would you give somebody who was new to the activity?*

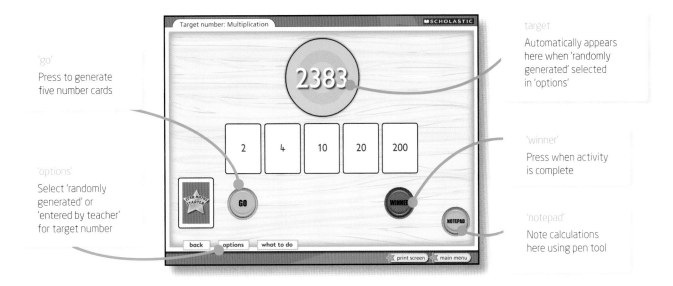

'go'
Press to generate five number cards

'options'
Select 'randomly generated' or 'entered by teacher' for target number

target
Automatically appears here when 'randomly generated' selected in 'options'

'winner'
Press when activity is complete

'notepad'
Note calculations here using pen tool

Finding reflections

Strand

Understanding shape

Learning objective

Visualise and draw on grids where a shape will be after reflection

Type of starter

Recall

Whiteboard tools

- Use 'options' to select the direction of the mirror line (horizontal, vertical or diagonal) and specify the size of the grid.
- Draw a shape in the white area. When complete, press 'done'. The white and shaded areas change places.
- Draw the reflection in the area that has just turned white.
- Press on a white or coloured square if you wish to change either the object or the image. When complete, press 'done'.
- Press 'check' to see if the reflection drawn is correct.
- Press 'reset' to clear grid and start again.

What to do

This activity offers a choice of reflection in the vertical axis, the horizontal axis, both vertical and horizontal axes, or a diagonal line. The object shape built may be as complicated as you or the children wish to make it – L-shapes, for example, are always interesting to experiment with. Invite children to come to the board to build the object and then to find the image.

Differentiation

Less confident: start with simple shapes in a vertical or horizontal line.
More confident: reflect more complex shapes in a diagonal line and with images one or two squares away from the mirror line.

Key questions

- *How do you know your shape or pattern is reflected correctly?*
- *Where would this shape be after reflection?*

'done'
- Press when finished drawing shape and reflection and the white area will move
- Press 'check' to find out if reflected shape is correct

'options'
- Select direction of mirror line
- Specify size of grid

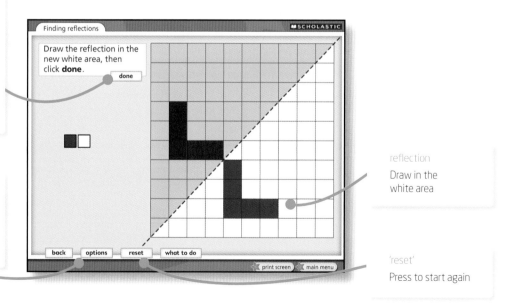

reflection
Draw in the white area

'reset'
Press to start again

Maps and directions

Whiteboard tools

- Press 'new' to generate a map with some paths blocked by roadworks and three objects to collect.
- Drag and drop the direction cards to prepare the route. Select one 'direction' card and one 'movement' card each time.
- Press 'move' to confirm the selection and move the recycling lorry along the route.
- Press 'show route' to display the directions selected so far.
- View the box at the top of the screen to identify which items the lorry collects along the route.

What to do

This activity requires the children to visualise where a recycling lorry will be after a translation and rotation; it develops their understanding of direction using mathematical language. Before the session, print the map grid from the CD-ROM, or photocopy it from page 46, as the children may find this useful.

From a given starting point, the children have to guide the lorry through the streets to the recycling centre. Instructions for direction are selected from: turn through 45° clockwise, 45° anticlockwise; 90° clockwise, 90° anticlockwise; 135° clockwise, 135° anticlockwise; 180°. The lorry may be moved one, two, three or four squares in any of these directions. There are, however, several different routes. At some junctions, but not all, the lorry can pick up items which are displayed on the screen - so the quickest route is not necessarily the best. The activity ends when the three items have been collected and the destination is reached. The position of the items changes randomly with each activity.

Differentiation

Less confident: ask the less confident children questions requiring only 90° or 180° directions at first.
More confident: ask the children for alternative ways of giving the same direction (for example, rotating through 45° clockwise).

Key questions

- *What is the most direct way to move from this point to that?*
- *Are there any alternative ways of moving from this point to that?*

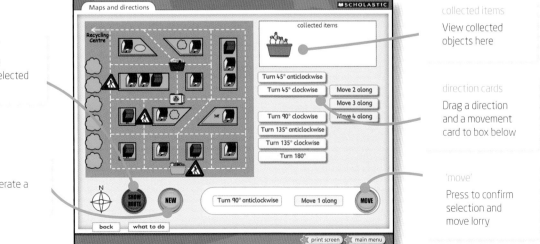

'show route'
Press to see directions selected so far

'new'
Press to generate a new map

collected items
View collected objects here

direction cards
Drag a direction and a movement card to box below

'move'
Press to confirm selection and move lorry

Finding rotations

Strand

Understanding shape

Learning objective

Visualise and draw on grids where a shape will be after rotation through 90° or 180° about its centre or one of its vertices

Type of starter

Refine

Whiteboard tools

- Use 'options' to select the direction and amount of rotation required, and specify the size of the grid.
- Draw a shape in the white area with either its vertex or its centre on the red spot. When complete, press 'done'. The white and shaded areas change places.
- Draw the rotation in the area that has just turned white. Press on a white or coloured square if you wish to change either the object or the image. When complete, press 'done'.
- Press 'check' to see if the rotation drawn is correct.
- Press 'reset' to clear grid and start again.

What to do

This activity offers a choice of rotation through 90° clockwise or anticlockwise or a rotation of 180°. The object shape built may be as complicated as you or the children wish to make it, with either a vertex or its centre on the red spot which is the centre of rotation. L-shapes are always interesting to experiment with. Invite children to come to the board to build the object and also, of course, to predict where the image will be when rotated.

Differentiation

Less confident: start with a rectangle rather than a square, as it is easier to see the rotation.

More confident: use more complex shapes than squares and rectangles. This software also enables rotations of shapes that do not have a vertex or their centre at the centre of rotation.

Key questions

- *Where is the image of each small square after the rotation?*
- *What would the rotation be if two rotations of 90° in the same direction were made consecutively?*

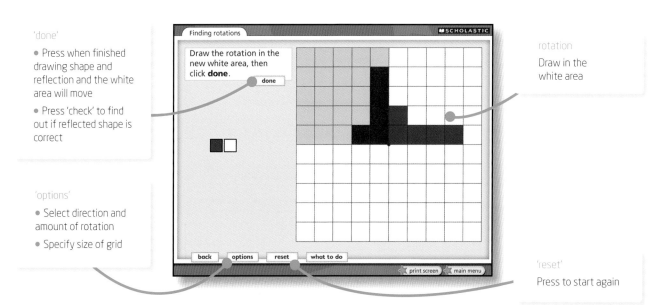

'done'
- Press when finished drawing shape and reflection and the white area will move
- Press 'check' to find out if reflected shape is correct

'options'
- Select direction and amount of rotation
- Specify size of grid

rotation
Draw in the white area

'reset'
Press to start again

Coordinates (ITP)

Strand

Understanding shape

Learning objective

Use coordinates in the first quadrant to locate shapes

Type of starter

Refine

Whiteboard tools

● Press the 'new marker' button to add an unlabelled marker to the grid. Use the arrows on either side to position it.
● Move the cross-hair over any marker and press the 'coordinates' button to show or hide the coordinates.
● Up to four quadrants can be used, if required.

What to do

The aim of this activity is to refine the strategies used by the children when identifying coordinates. The ITP automatically opens in a single quadrant. To make a new marker, select the icon marked 'X' and press 'new'; then use the arrows in the box above 'new' to position the marker on the grid. Repeat until you have created a series of marks. The marks must be placed one at a time, or they will pile up on top of one another.

Briefly revise with the children how to write coordinates. Then ask them, using their individual whiteboards, to identify the coordinates of the markers. Next, open the 'cross-hair' icon on the far right of the screen and turn on the 'coordinates' marker, which appears below it. The cross-hairs can be manually placed anywhere on the grid and will automatically reveal its current coordinates. Check results.

Differentiation

Less confident: place a number of markers on either the vertical or horizontal axis so that when the children are identifying the coordinate they can see the relationship between the x and y axis numbers.

More confident: by pressing the button that appears to the left of the coordinates icon, introduce the more confident children to coordinates within two or four quadrants. Investigate how coordinates change across the four quadrants.

Key questions

● *Where else would we commonly find coordinates?* (Consider Ordnance Survey or A–Z maps, online mapping systems such as Multimap or Google Maps, and games like Battleships.)
● *Can you recommend any strategies or think of any tips that we could use to help us remember the order of numbers in a pair of coordinates?* (For example, 'X is a cross' [across].)

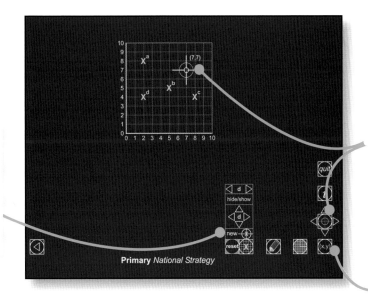

'cross-hair'

Press over a marker to focus on the coordinates

'new marker'

● Press to create new marker

● Can either be placed manually around the grid or moved using the arrows

'coordinates'

Can be hidden or revealed wherever the cross-hair is placed

Fixing points (ITP)

Strand

Understanding shape

Learning objective

Estimate angles and use a protractor to measure them

Type of starter

Refine

Whiteboard tools
- Use the 'grid size' menu to specify the size of the grid.
- Press on any number of points on the grid to create a polygon.
- Use the 'protractor' to measure the angles.
- Press the 'angles display' button to check answers.
- Press 'reset' to remove any lines drawn.

What to do

The aim of this activity is to help the children refine their skills in measuring angles and distance. The ITP opens as a blank grid. Pressing on a series of dots will allow you to create a simple line shape and each dot that you select is marked by a sequential letter. You complete the shape by returning to dot 'A'. Press either the 'ruler' or 'protractor' button to launch these tools; both can be dragged, dropped and rotated to make measurements, and can be used on the screen at the same time. Once you have got used to the program, you can change the style and size of the grid in order to make the accurate measurement of angles and distances more challenging.

Differentiation

Less confident: start by making one angle between two markers. Ask children to come to the board and use the protractor to measure the angle.
More confident: create a shape with several points and reveal the angles. Challenge the children to find where each angle is, after hiding the letters.

Key questions
- *How do we position a protractor in order to correctly measure an angle?*
- *Once we have identified the distance between two dots, do we need to use a ruler?*

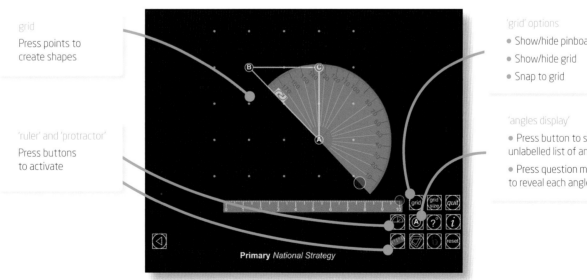

grid
Press points to create shapes

'ruler' and 'protractor'
Press buttons to activate

'grid' options
- Show/hide pinboard
- Show/hide grid
- Snap to grid

'angles display'
- Press button to show unlabelled list of angles
- Press question marks to reveal each angle

Calculating angles (ITP)

Strand

Understanding shape

Learning objective

Calculate angles in a triangle or around a point

Type of starter

Refine

Whiteboard tools

● Press the 'angle' button and select the 180° angle.
● Press the 'shape' button to select a shape. This will appear on the 180° angle, creating two unmarked angles.
● Press the 'colour' button to change the colour of the shapes.
● Use the 'protractor' to measure the angles.
● Press the question marks to reveal the angles.

What to do

This activity aims to refine the children's skills in measuring an angle, either within a shape or on a 180° line. Select the 180° angle from the 'angle' menu. Then, select a single shape from the 'shape' menu and ask the class how they would measure either the angle within the shape or the outside angle: both are marked with a question mark. Ask: *What information would we need?* Reveal one of the angles by pressing the question mark and ask: *How could we find the size of the other angle?* Take any answers and then, before revealing the answer, measure the angle using the on-screen protractor. Repeat this with other shapes, or use multiples of a shape by pressing the 'shape' button twice to produce two shapes on one line - thereby providing more angles to measure.

Differentiation

Less confident: consolidate measuring an angle round a point using a single shape on a 180° base only.

More confident: select the top item under the 'angle' button, which shows a circle (360°). Begin measuring and estimating angles within this. Ask: *How could we estimate the size of an angle round a point, using these shapes?*

Key questions

● *When multiples of a regular shape are used* (for example, an equilateral triangle), *what happens to the outside angle - is there a pattern?*
● *Are there any shapes we already know the angles of? If so, what are they?*

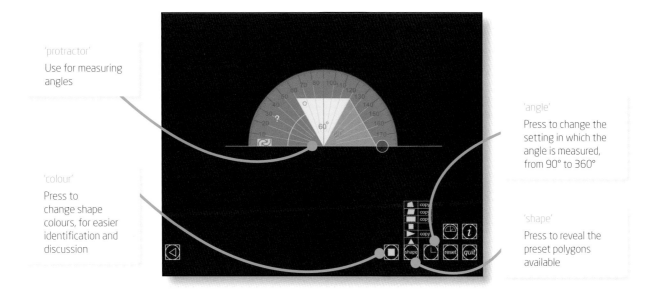

'protractor'
Use for measuring angles

'colour'
Press to change shape colours, for easier identification and discussion

'angle'
Press to change the setting in which the angle is measured, from 90° to 360°

'shape'
Press to reveal the preset polygons available

Convert the weight

Strand	Whiteboard tools

Strand

Measuring

Learning objective

Select and use standard metric units of measure and convert between units using decimals to two places

Type of starter

Read

Whiteboard tools

- Drag and drop items on the left of the screen into the pan or add some preset weights.
- Turn the digital readout on or off on the analogue scale, as required.
- Select 'options' to make changes to the divisions on the face of the scales.

What to do

The aim of this activity is for the children to learn to accurately read and convert kilograms to grams, or vice versa. Drag items into the pan and ask the children (working in pairs) to write down the weight on their individual whiteboards – using decimals and the notation that is opposite to that used on the screen (for example, grams if kilograms are shown). Then tell them to show their partners. Use the digital readout to reveal exact weights. Assess errors and ask: *How can the face of the scales help you when you are reading the weight?*

Repeatedly select and drag different food items and preset weights into the scales' pan. The children should get faster at reading the weight.

Differentiation

Less confident: use the 'options' menu to increase the number of subdivisions on the face of the scales to 10.

More confident: record the initial weight of the food items, then remove two items and record the new weight. Ask: *What is the difference between the two weights?*

Key questions

- *How can we use the weight of the bag of flour to help estimate the weight of other products? What other food products would be useful guides?*
- *What is the total weight of these two items in grams and kilograms?*

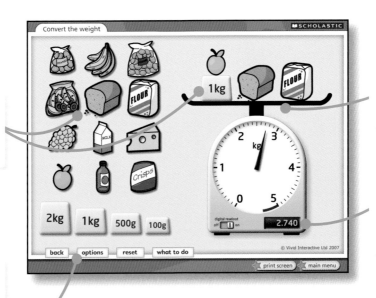

items and weights
Drag food items or preset weights onto scales' pan

scale's pan
Add or remove items by dragging them off or on

'digital readout'
Can be turned off or on, on analogue scale, to reveal exact weight of items

'options'
Set maximum weight, subdivision, and whether or not to show numbers on analogue scales

Measuring jug: read scales

Strand

Measuring

Learning objective

Use standard metric units of measure and convert between units using decimals

Type of starter

Rehearse

Whiteboard tools

- Press the 'options' button to set the scale and sub-divisions on the jug and also select the fill steps you require.
- Press the 'in' switch to fill the jug, and press it again to stop filling.
- Press the 'out' switch to empty the jug, and press it again to stop emptying.
- Press 'reset' to start again.

What to do

Ask the children what they can tell you about the scale on the jug and what the smaller sub-divisions mean. Fill the jug part way up to a certain level; the level will sit either on the small sub-division mark or halfway between the sub-divisions. Then ask the children to give you the reading, first in ml and then in litres, using decimals (for example, 8800ml and 8.8 litres). By filling and emptying the jug, the difference between two levels can be found.

Differentiation

Less confident: in the early stages, ask the children to read the scale to the nearest 200ml mark.

More confident: challenge these children to give a more accurate reading.

Key questions

- *I have filled the container to 6200ml. What is that in litres?*
- *A garage orders 20,000 litres of petrol. It sells an average of 1250 litres per day. How long will the supply last?*

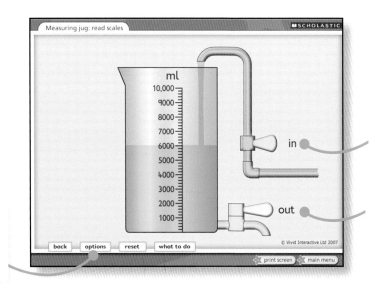

Calculating perimeter

Whiteboard tools

- Use the 'options' menu to specify the size of the grid.
- Use the highlighters to build up shapes on the grid.
- Press on a white or coloured square if you wish to change the shape.
- Press 'clear' to start again.

What to do

The object of this activity is to demonstrate to the children that there is a quicker
way of calculating the perimeter than just finding the distance all the way round. The
whiteboard tool may be used to build up rectangles of different sizes and shapes.
Rather than tell the children that the perimeter is the sum of twice the length plus
twice the width, see if they can discover this for themselves. Assume that each small
square is 1 × 1cm so that the perimeter is measured in cm.

An additional activity involves building different shapes - of 12 small squares,
for example. Investigate the areas of the different shapes formed and then check
whether the perimeter stays the same.

Differentiation

Less confident: to begin with, allow the children to find the length all the way round
the rectangle, to ensure that they understand the concept of perimeter.
More confident: ask the children to think of ways in which they could find the
perimeter more quickly than adding lengths all the way round the shape.

Key questions

- *How do you think you could calculate more quickly the perimeter of a rectangle?*
- *If the perimeter is twice the length plus twice the width, is there another way
you could make the same calculation?*

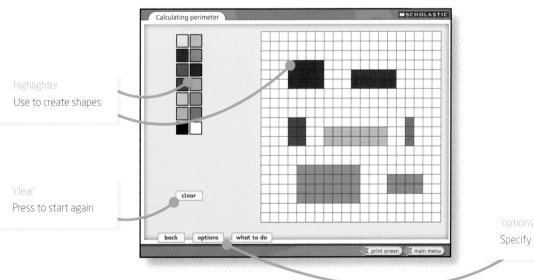

highlighter
Use to create shapes

'clear'
Press to start again

'options'
Specify size of grid

Finding area

Strand

Measuring

Learning objective

Calculate the area of rectilinear shapes

Type of starter

Refine

Whiteboard tools
- Use the 'options' menu to specify the size of the grid.
- Use the highlighters to build up shapes on the grid.
- Press on a white or coloured square if you wish to change the shape.
- Press 'clear' to start again.

What to do

The objective is to challenge the children to find the area of rectangles without counting squares individually. An interesting activity is to build different rectangles (each using 12 small squares, for example) and investigate the areas of the different shapes formed. Assume that each small square is a 1 × 1cm, so that the area is measured in cm². Figures of various other shapes and sizes can also be formed using the small squares.

Differentiation

Less confident: start with simple rectangles until the children get used to calculating the area, which can then be checked by counting the small squares.
More confident: make up some non-rectangular shapes using the small squares, and ask the children to investigate their areas. Challenge them to find a way of measuring the areas without counting the small squares.

Key questions

- *For the shapes drawn here, how do you think you could calculate the area without counting the small squares?*
- *Can you explain why calculating the length times the width gives the area of a rectangle?*

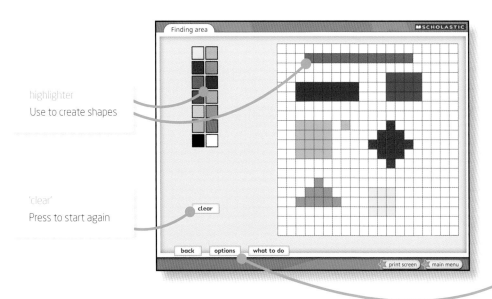

highlighter
Use to create shapes

'clear'
Press to start again

'options'
Specify size of grid

Area (ITP)

Strand

Measuring

Learning objective

Estimate the area of an irregular shape by counting squares

Type of starter

Rehearse

Whiteboard tools

● Specify the size of the grid required using the 'grid size' menu.
● Press the 'pinboard' button to change the grid to a pinboard, if required.
● Press the 'shape colour' button to select from a range of colours for solid shapes.
● Use the 'shape control' menu to select a solid shape to automatically add to one square space on the grid.
● Press the 'reset' button to clear solid shapes from the grid.
● Use the 'elastic band' menu to access a virtual elastic band that can be stretched around the corners of the grid.

What to do

The aim of this starter is to rehearse the skills and strategies needed to find the perimeter of a polygon. Below are examples of how polygons can be created, using either the automatic solid shapes or elastic bands. Once a shape is created, review how to find its perimeter and what strategies the children might use. The level of complexity and challenge can be altered, particularly when using the elastic bands, to quickly develop and test individual or group theories.

Differentiation

Less confident: use automatic solid shapes to create regular polygons. Adding one triangle at a time, demonstrate how the perimeter increases.
More confident: use the elastic bands to create more challenging polygons.

Key questions

● *If we didn't have a ruler, how could we estimate the size of this shape?*
● *Does it make any difference if the shape is made out of elastic bands or solid shapes?*

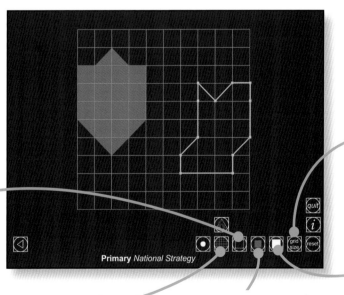

'elastic band' menu

● Select number of points for elastic band

● Select up to two elastic bands at a time

● Press buttons to erase shapes drawn

'grid size' menu

Select size required

'shape control' menu

Press to add shape to one square space on grid

'grid/pin board'

Changes grid to pinboard

'shape colour' button

Offers a range of colours

Primary *National Strategy*

Data handling (ITP)

Handling data

Learning objective

Solve problems by collecting, selecting, processing, presenting and interpreting data, using ICT where appropriate

Type of starter

Read

Whiteboard tools

- Press the 'data' button to reveal preset data or to create your own.
- Select a chart type for the data by pressing one of the 'chart buttons'.
- Choose to 'hide/show' elements of the data table, including the colour key, titles, numbers and percentages.
- Use the 'increase/decrease' buttons to manually change bar chart data.
- Use the 'max' button to type in the maximum value on the numerical axis of a bar chart when creating your own data. The input number range is 0.9 to 90,000.

What to do

The aim of this activity is to present preset data to the class so that the children can analyse and interpret the information it provides. You may find it helpful to begin with the preset data on eye colour, accessed via the 'data' button. Try presenting two chart types at once, so that they can be compared. To do this, press the 'Esc' button, which will minimise your first screen. Then press on the ITP to launch a second version of the same program (see notes on page 6 of this book to learn more). It is worth analysing the categories and data for each eye colour; then identify other colours that could be present. Find out whether the balance between the categories reflects the variations shown within the class. Using the 'increase/decrease' buttons, adapt the bar chart so that it shows the class results. Ask: *Can comparisons with the preset data be made? What are the differences and why do we think they exist?*

Differentiation

Less confident: encourage the children to read the bars using just the scale. Hide the numbers and reveal them only at the end of the activity.
More confident: view the data as a pie chart. Shown in this way, ask: *What general statements can be made about the data? Could the data be translated into fractions?*

Key questions

- *What statements about the data can you make before reading any figures?*
- *Can you identify any equivalent fractions?*

'increase/decrease'
Change data manually

'hide/show'
Hide or show elements of the data

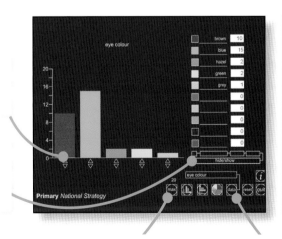

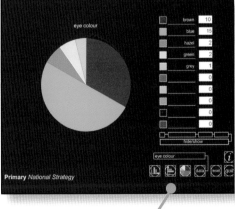

'max'
When creating new data, type above this button the data range to be shown; then press 'max'

'data'
Press to reveal preset data or to create new data

three 'chart buttons'
Choose chart type

Line graph (ITP)

Strand

Handling data

Learning objective

Interpret line graphs

Type of starter

Read

Whiteboard tools

- Use the 'arrows' to alter the numbers of units along the *x* axis.
- Press the 'data' button for a list of preset line graphs available.
- 'Hide/show' the data table, as you prefer.
- The 'reset' and 'line graph' buttons reset the data in the table.

What to do

The aim of this starter is to improve the children's skills in reading line graphs and extracting the general information that they present. This ITP opens as a blank graph and table. Four preset charts can be accessed by pressing the 'data' button. All information can be edited or altered, by either dragging the red squares on the line graph up or down, or by editing the data in the right-hand side of the table. In the case of the image below, 'journey' has been selected using the 'data' button. Direct the children as they analyse the data relating to the lorry driver's journey, highlighting the importance of the *x* and *y* axes. The other line graphs available within this ITP show that the *x* axis is often used to plot time, but that it can be used differently (see the 'heights' graph).

Differentiation

Less confident: talk through the line graph, asking the children: *What is the story that this line graph tells?*
More confident: examine trends within the line graph and make predictions as to how the graph could continue. Ask: *What might the next day's line graph look like?*

Key questions

- *What does the information on the line graph represent and how should we analyse it?*
- *What 'story' is the information telling us?* (For example: *How can we tell what has happened on the lorry driver's journey?*)

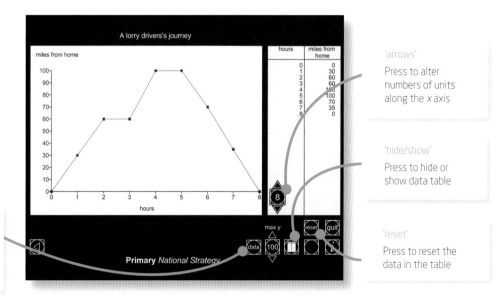

Fairground ride

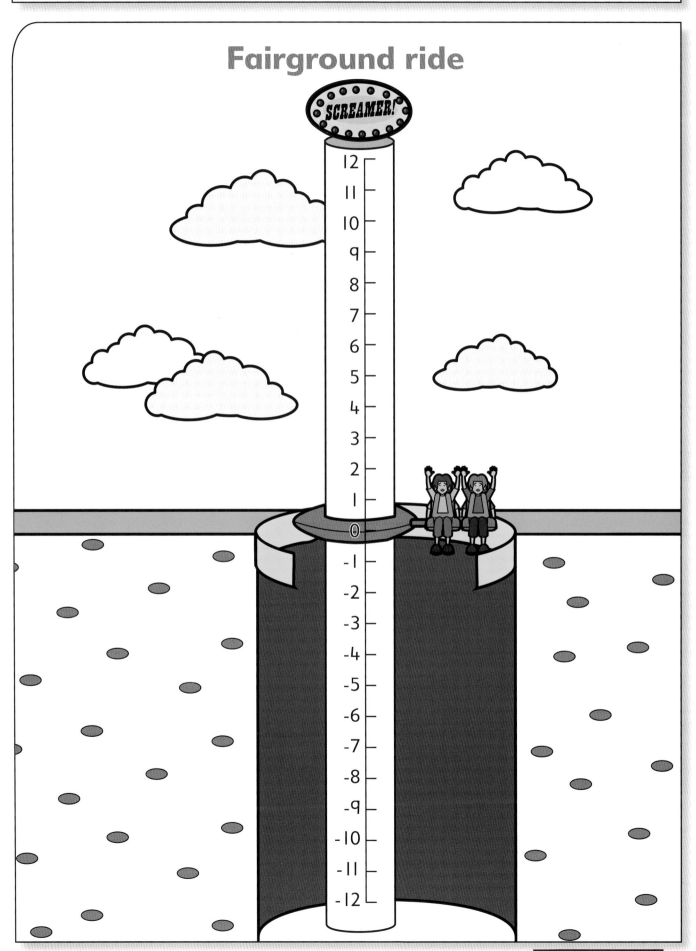

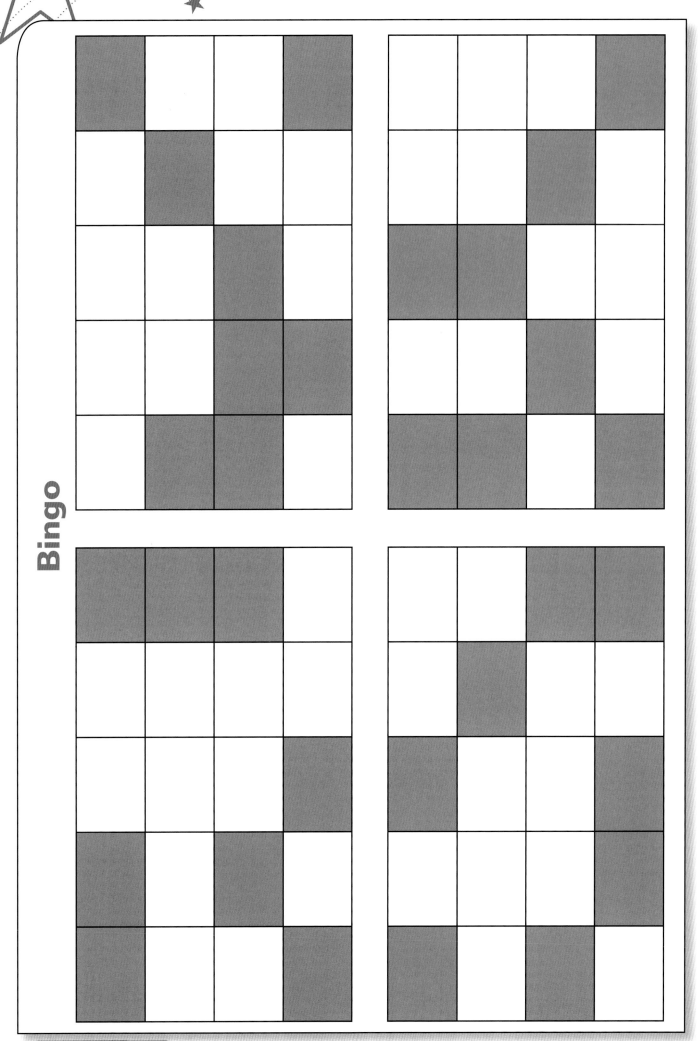

Bingo

Target number: multiplication

🔖 Find the target numbers using the cards below.

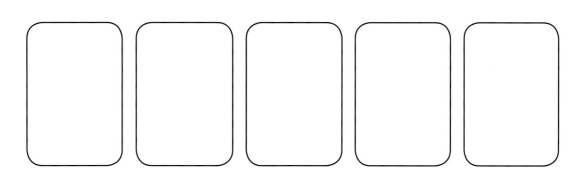

🔖 How I worked out the target answer.

_____

_____

_____

_____

_____

_____

_____

_____

Maps and directions

📖 Plan your route using the map below.

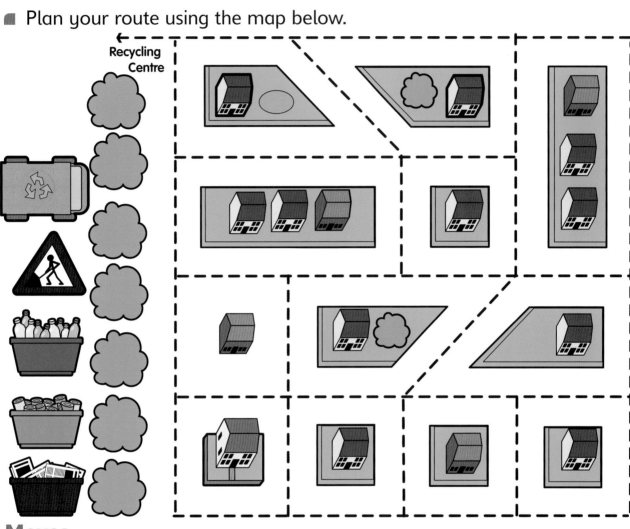

Recycling Centre

Moves

_____

_____

_____

_____

_____

_____

_____

Star Maths Starters diary page

Name of Star Starter	PNS objectives covered	How was activity used	Date activity was used

Also available in this series:

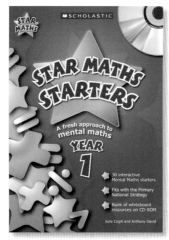

ISBN 978-1407-10007-4

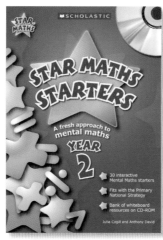

ISBN 978-1407-10008-1

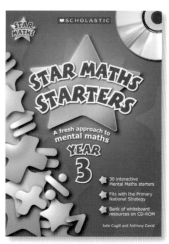

ISBN 978-1407-10009-8

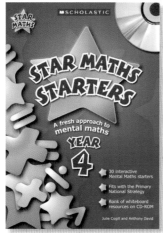

ISBN 978-1407-10010-4

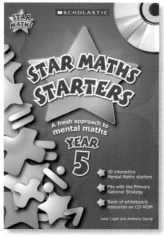

ISBN 978-1407-10011-1

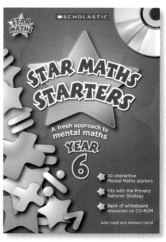

ISBN 978-1407-10012-8

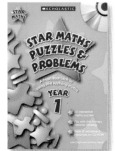

ISBN 978-1407-10031-9

ISBN 978-1407-10032-6

ISBN 978-1407-10033-3

ISBN 978-1407-10034-0

ISBN 978-1407-10035-7

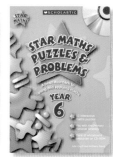

ISBN 978-1407-10036-4

To find out more, call: 0845 603 9091
or visit our website www.scholastic.co.uk